This book is to be returned on or before the last date stamped below.

301.16

070.1

T.P. RILEY COMMUNITY SCHOOL

LIBREX

003130

media watch
tv and video
Bosie Vincent

Wayland

media watch

**Advertising
Newspapers
Magazines
TV and Video**

Series Editor: James Kerr
Designer: John Christopher

Consultant: Julian Bowker,
Education Officer, British Film Institute.

Front cover picture: Popular game show *Wheel of Fortune*.

First published in 1993 by
Wayland (Publishers) Ltd
61 Western Road, Hove
East Sussex, BN3 1JD

© Copyright 1993 Wayland (Publishers) Ltd

British Library Cataloguing in Publication Data
Vincent, Bosie
TV and Video. - (Media Watch Series)
I. Title II. Series
791.45

ISBN 0-7502-0729-9

Typeset by Strong Silent Type.
Printed and bound in Spain by Graficas Estella.

Picture Acknowledgements
The publisher would like to thank the following for providing the pictures used in this book: Allsport/Adrian Murrell 34; Aquarius Library 18; BBC 5,8,15,16,17,19,20,21,23; Channel 4 Stills 11,22,30; Eye Ubiquitous COVER (Paul Seheult),36,37,38 (Paul Seheult),39 (Paul Seheult),41 (Matthew McKee),44; Kobal 7,9,10; London Features International/Ron Wolfson 35; Popperfoto 4,26,28-9; Rex Features Ltd. 12-13; Topham 6,25,32,40,42-3.
The artwork is by John Christopher.
Where necessary, the publishers have attempted to gain permission from the artists who feature in the photographs used in this book. The publishers apologize if there have been any oversights.

media **watch**

contents

At the touch of a button: the history of TV	4
Soaps, sitcoms and serials: TV genre	10
From script to screen: TV production	16
Who calls the shots? Ownership and control	22
Telly addicts: the audience	30
Tuning in to the future: new technology	36
Glossary	45
Further reading	46
Notes for teachers	47
Index	47

tv and video

At the touch of a button:
the history of TV

There are few, if any, countries in the world which have been untouched by the massive impact of television. It has revolutionized the way we communicate and gather information and has also helped us to learn about each other. TV could be said to have made the world a smaller place. One hundred years ago it would not have seemed possible for people to sit in their own homes watching live, moving pictures from far-flung corners of the earth, or even the moon!

Nowadays, it is difficult to imagine a world without television but TV has only been with us for about half a century. Video, which many of us take for granted, has been around for even less time.

John Logie Baird, the Scottish engineer, is generally credited with the invention of television, although there have been rival claims from scientists in Russia and Spain. Baird gave his first successful demonstration of TV in 1926 when he transmitted the image of a boy's face.

Ten years later, on 2 November 1936, the British Broadcasting Corporation (BBC) began its first regular television broadcasts from Alexandra Palace in North London. To begin with, only people living within a 50 km radius of the transmitter could pick up the signal. Very few people owned a television set, since they cost the equivalent of a small car.

LEFT Watching television became a popular family activity from the 1950s onwards.

BELOW In the early days of radio and television, presenters wore very formal clothing.

media **watch**

ABOVE
Colour television was not introduced in Britain until 1967. This 1957 magazine picture shows readers what TV star Bob Monkhouse looks like in colour.

Output of television programmes increased when EMI's Emitron system, which used an electronic scanner, replaced Baird's clumsy mechanical scanner. However, all programmes had to be broadcast live because it was not yet possible to record shows and broadcast them later on.

Regular broadcasting services began in the USA and Germany in 1939. The BBC closed down their service on 3 September 1939 when Britain declared war with Germany. It was feared that German planes might use the transmission signals as homing devices during bombing raids on London.

After the end of the Second World War in 1945, the television industry grew rapidly. By 1952 New York and Los Angeles both had seven TV stations; in Britain 80% of the population was within range of a transmitter. The BBC's coverage of the coronation of Queen Elizabeth II in 1953 was a massive boost for television. It was watched in Britain by an estimated 20 million people on only 2.5 million sets, although many more were subsequently sold as a result of that broadcast.

By 1955, when Independent Television (ITV) began in Britain, cinema audiences had plummeted to one third of what they had been in 1950. This was mainly due to the competition from

tv and video

television. Around the mid-1950s, the first video recorders began to be used by television companies. This meant that television programmes could be filmed, edited and then broadcast at a later date.

The first truly world-wide television event was the final of the 1966 World Cup at Wembley Stadium in London. It was covered jointly by the BBC and ITV, using forty-five cameras. The black and white pictures were broadcast by satellite around the world to over 400 million people.

The first colour TV broadcast was of tennis from Wimbledon in 1967. It was shown on the second BBC channel, BBC2, which had only started operating three years earlier. The introduction of colour was a further boost to the popularity of television.

In the 1960s and 1970s, television spread to many other countries throughout the world. People began to realise that TV could provide information and education as well as entertainment. In developing countries in Africa, for example, TV is used to promote both education and health. There are programmes giving instruction and advice on topics like family planning. Television also plays an important part in political elections, allowing opposing parties to broadcast their messages to the nation. In a country like India, which has the largest population of any democracy (a country where everyone has a vote), television has a vital role to play in ensuring that everyone can make an informed choice.

BELOW
The Honeymooners **was a hugely popular American sitcom in the 1950s. It is still repeated today.**

media **watch**

Today we live in a global village – a world linked by the shared experience of television. The use of satellites to beam television pictures instantly around the world means that someone in Britain can watch the same programme at exactly the same time as someone in Australia. Television coverage of international events such as the Olympics, the World Cup or the Live Aid concert of 1985 allows hundreds of millions of people around the world to watch the same thing at the same time. TV is the only medium which can focus the world's attention in this way.

Some people believe that too much television and video viewing can be bad for you. They argue that it encourages people to be intellectually passive, in other words to sit back and not think for themselves – to turn into 'couch potatoes'. However, studies have shown that Finland has one of the highest levels of literacy and yet the children also have the highest consumption of television viewing.

Another fear is that television programmes which contain scenes of violence have a harmful effect on viewers. Scientists have estimated that in the USA the average young person will have witnessed 15,000 murders on television by the age of eighteen. However it is very difficult to prove that these programmes actually have a damaging effect on viewers.

It is also argued that the people who control television are in a position to control their audience. For example, when a government controls all of the

BELOW
**A historic TV event –
Abba win the 1974
Eurovision Song
Contest.**

tv and video

ABOVE During the 1970s, many people were concerned about the amount of violence in the American cop show *Starsky and Hutch*.

television channels in a country, as is the case in China, they can be very selective about what they show. They may only show things which make the government look good. They are in a position to prevent people who oppose the government from expressing their views on TV – a practice known as censorship. By controlling television, it is possible to control what information is available to viewers and to present a biased viewpoint.

No comprehensive study of modern society would be complete without a look at the impact of television and related technologies such as video. The following chapters of this book will examine how television is created by programme makers and consumed by audiences, and will also look ahead to what we might expect television and video to be like in the future.

Activities

• Compile a survey on people's television viewing habits. Ask your friends, parents and teachers how many hours they spend a week watching television programmes or videos. Compare this with time spent on other activities such as the following:
- Reading
- Listening to the radio, tapes or CDs
- Computer games
- Outdoor games and sports
- Homework

• Divide these activities into 'media' and 'non-media'. For example, TV, video and radio would be 'media', whereas playing football or doing your homework would be 'non-media'. How many hours are spent on each category? Which activity is the most popular overall?

9

media **watch**

Soaps, sitcoms and serials: TV genre

The programmes we watch on television can be divided into different types or 'genres'. For example, children's television could be described as a genre. Other examples are soap opera, situation comedy, drama and current affairs.

Each genre has its own recognizable style. We can instantly tell the difference between a game show and a serious news programme just by the appearance of the presenter and the set. There are lots of other things which make up the style of a specific genre: the theme music, the way the presenters speak, the way the actors act and the way in which ordinary people are represented.

BELOW
Cartoons are a TV genre. One of the key ingredients in this genre is the catchphrase, repeated by the main character a number of times in each show, for example Fred's 'Yaba Daba Doo!'

Some types of programme are much cheaper to make than others. A game show, for example, can be made for about £80,000 whereas a costume drama may cost ten times as much. Game shows require one set for a whole series and are usually hosted by just one presenter. The script hardly varies from week to week and even the big prizes cost relatively little in terms of big television budgets. Game shows are also very popular. The fact that they are cheap to produce and attract large audiences accounts for the large number of game shows on our screens.

People have different expectations from different genres. We expect simply to be amused by a comedy programme whereas we may watch a documentary or a nature programme in the hope of finding out about something.

Programme makers carefully follow the formula of each genre. If viewers can instantly place a new programme within a certain popular genre, then they may tune in and watch it simply because they normally like that type of programme. However, some people argue that making programmes to a formula in this way is repetitive and unchallenging for the viewer. Let's look at a few different genres in more detail.

ABOVE
Roseanne Barr and the cast of *Roseanne* celebrate the 100th episode of the American situation comedy. *Roseanne* is popular because it does not simply rely on 'jokes'. It finds humour in the day-to-day life of the Connor family. In doing so, it deals with problems such as adolescence, relationships and unemployment.

Soap Opera

Soap operas originated in America as radio dramas sponsored by soap powder companies. They are continuous or long-running drama series which usually take place in a particular area or location such as a street, a pub or a family house. The scenes are short and the action is fast-moving. Different problems are always cropping up and the plots are usually complicated. The acting is often exaggerated or over-the-top. Each episode ends with a cliffhanger so that the audience is left wondering what is going to happen next. This encourages them to tune in to the next episode. *Neighbours* is a good example of a soap.

Situation Comedy

Like soap operas, sitcoms are usually set in one place. The situation remains the same throughout the series, and the comedy and the story-lines arise from it. For example, the 'situation' in *Cheers* is a Boston bar with a small group of regulars who, in spite of the fact that they frequently tease and insult each other, are good friends. Each week something may arise which threatens to change this situation, but by the end of the episode, everything has usually returned to the way it was.

Sitcoms are very distinctive because they are often recorded in front of an audience so that the viewer hears the audience laughter throughout. This is to encourage the audience at home to laugh. Many children's cartoons also use 'canned' (recorded) laughter in this way.

media **watch**

News and Current Affairs

These programmes deal with fact, not fiction, and usually concentrate on serious political and social issues. Therefore the style of this genre is very formal and serious, from the opening title music to the appearance and manner of the presenters. We watch these programmes to be informed rather than to be entertained. Because of the style of this genre, viewers may tend to accept as fact information given by the presenters – they have an authoritative tone of voice and way of speaking. The trust that the audience has in what a news-reader says is so powerful that the broadcasting authorities in some countries have banned news-readers from appearing in adverts. People tend not to question what news-readers say and this makes them different from other television presenters.

RIGHT TV images direct from the frontline of a war zone tend to reinforce our belief in what we hear and see on news programmes.

The style of local news programmes varies from that of national and international TV news. Local news programmes tend to include more light-hearted news stories and the presenters often chat with each other in an informal way.

media **watch**

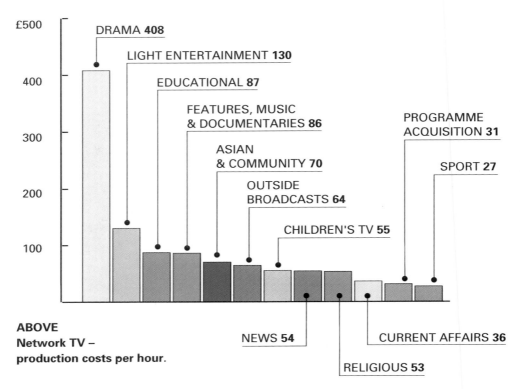

ABOVE
Network TV – production costs per hour.

Game Shows

Most game shows have the following formula. A host – usually male – introduces us to the contestants, who are ordinary members of the public, although they are often accompanied by celebrities. The host explains the rules of the game and then shows us the prizes on offer: maybe cash, a holiday or a range of household appliances. The game is then played. The contestants are gradually eliminated until the final round, where a single contestant or a couple will try to win the big prize. The excitement of the programme builds up to this moment, although the big prize is not necessarily won each day or week. This ensures viewers will tune in to the next show.

The game show genre is typified by bright, garish sets and enthusiastic presentation. Cheerful music is played at intervals throughout the programme, such as after each round of the game. The host is sometimes accompanied by glamorous assistants, usually female.

In game shows, members of the public are presented to us in a very different way from the way in which they are seen in news and current affairs programmes. They are encouraged to behave in a cheerful, carefree and rather artificial manner. However, some alternative game shows, such as *The Crystal Maze*, could be said to bring out genuine fear, tension and excitement in the contestants.

tv and video

Activities

- Imagine that you are making a programme called *The Day the School Burned Down*. Look at the following examples of different genres:
  - Horror
  - Documentary
  - Hospital Drama
  - Situation Comedy
- How would your programme vary if it were made in each of these genres? Think about differences in the music, the set, the script and the style of the actors or presenters.

RIGHT
Bruce Forsyth is a true TV personality. His popularity – based on his delivery of jokes and catchphrases and ability to have some fun at the contestants' expense without making them look too stupid – was a key part of the success of BBC's *The Generation Game* in the 1970s. The show has been revived in the 1990s. Hosted by Bruce and Rosemarie Ford, it uses virtually the same formula as before.

media **watch**

From script to screen:
TV production

Different television programmes are made in different ways. A sports programme might be broadcast directly from a sports ground, whereas a quiz show is usually filmed in a studio in front of an audience. A television drama might be filmed in different locations throughout the country or made entirely on studio sets.

BELOW The amount of money spent on costumes for a show like *The House of Eliott* will have been worked out before the series goes into production.

Many TV programmes are produced by people who work for a TV channel. However, some programmes are produced by independent companies and are then sold to TV channels for a fee. This chapter describes the different stages of production that a studio-based programme produced by a TV channel has to go through before it is broadcast. There are three main stages; pre-production, production and post-production.

Pre-production

The head of the TV channel and the heads of department will decide which scripts or ideas for programmes can be turned into successful TV shows. For example, the head of the channel and the head of the light entertainment department may decide that they like an outline of a sitcom. They will then commission the programme. This means they appoint a producer and tell him or her what the budget – the amount of money the channel intends to spend on the programme – will be.

RIGHT A television studio.

media **watch**

The producer must co-ordinate the making of the programme from start to finish. He or she is responsible for the content of the programme and must try to make sure that it does not go over budget during production.

If the show features a star then he or she may have a great deal of influence during the production process. Noel Edmonds, for example, has a large say in deciding what the content of *Noel's House Party* will be. If a star and a producer cannot work together then the producer is often replaced, since it is the star who is going to make people want to watch the programme.

It is not unusual for a star to produce his or her own show. Bill Cosby and Oprah Winfrey own TV production companies. They produce and star in shows and then take a large slice of the fee when the show is sold to a TV station.

One of the producer's first tasks is to appoint a script editor, who in turn appoints writers. Comedy programmes such as *Hale and Pace* or *Alas Smith and Jones* may use over fifty different writers during a single series. The script editor often accumulates about three times more written material than is actually needed for each show. Working with the producer, he or she has to edit this down to the required amount.

BELOW Designers build models of sets to find the right 'look' for a show.

tv and video

ABOVE A TV camera operator in a studio.

Several months before a programme is recorded, the producer appoints a designer. He or she is in charge of creating the 'look' of the programme. Designers design and co-ordinate the building of sets and also advise the props and costume departments.

Designing and building sets is one of the most expensive parts of television production, so producers are keen for the writers to reuse the same set for several scenes in order to save money. The producer will appoint a composer to write any original music that will be used in the show, such as a theme tune.

One of the producer's most important tasks is to appoint a director, who is responsible for directing the actors and the film crew. The director must break the script down into separate shots and then decide how each shot is to be staged. In other words, he or she decides where to position the cameras and where the actors will move. The director must also give the film crew and the actors instructions on what style of filming or acting each shot requires.

Some television programmes, usually dramas, require the preparation of a storyboard before filming takes place. This is done by the director or a specially appointed storyboard artist.

19

media watch

A storyboard is a set of pictures, like a comic strip, which show what each shot will look like. The pictures indicate the framing, composition and camera angles of each shot. They also give information about the lighting, such as which parts of the frame will be brightly lit and which parts will be in shadow.

Rehearsals for a television programme, unlike those for a stage play, usually start only days before the programme is recorded. The director and actors work in a rehearsal room with a plan of each set marked out on the floor with sticky tape.

The director will go through the script with the performers, who will practise the delivery of the lines, the timing of the jokes and so on. Dummy props are used, in case the real ones get broken.

The writers may have to make adjustments to the script at this stage, so they need to be present at the rehearsals.

Production

At the end of the rehearsal week, the programme is recorded on video tape in the studio. In the case of a sitcom or quiz show, the programme is often filmed before a live audience so that laughter and applause can be recorded at the same time. This is very similar to an audience watching a play in a theatre except that the studio is full of large television cameras. This means that the audience do not always get a clear view. They occasionally have to watch on televisions, called monitors, above their heads.

If something goes wrong or an actor makes a mistake, a retake will have to be done. The actors must go through the scene again and the camera crew will re-record it. Sometimes it is necessary to do a lot of retakes. If a programme is being broadcast live, this is not possible and the mistakes will not be edited out. Live broadcasts require meticulous pre-production so that they will run smoothly.

BELOW Filming on location.

tv and video

Some parts of a programme may have to be filmed outside on location. These scenes are usually filmed together, over two or three days, before the rest of the series is filmed in the studio. This means that the scenes filmed on location can be shown on the monitors at the appropriate moments during the studio recording.

Post-production

Editing is the putting together of the different scenes that have been filmed, to make a whole programme. Video editing involves recording scenes from one video tape to another, perhaps removing unwanted shots or rearranging the order of scenes.

The editor must be sure that the programme runs to the correct length so that it can fit accurately into the schedules. The editor will also add the composer's music to the programme.

When the programme has been completed it is ready to be broadcast. It is played on a video machine and simultaneously transmitted in the form of an electronic signal from the TV station to powerful transmitters all over the country. These relay the signal to people's houses via their TV aerials.

ABOVE
TV production crew sit in front of a bank of screens when a programme is being transmitted, so that they can monitor what is being broadcast. On shows which are going out live, they look at the pictures from a number of different cameras and make sure that the best images are being broadcast.

Activities

- Prepare a storyboard for a drama called *The Liar,* set in your school or local youth club. Sketch each individual shot indicating the type of framing – long shot, medium shot or close-up.
- Choose a range of locations for the drama to take place in. For example, on the school field or in the dinner hall.
- Under each picture, write down the dialogue that will be spoken during the shot and choose appropriate music to go with it. Produce costume and set designs to go with your storyboard.

media **watch**

Who calls the shots?
ownership and control

Have you ever thought about who is actually in control of the television channels that you watch, and the kinds of decisions these people have to make? Who chooses what type of programmes get made and broadcast? Who decides what programmes we are and aren't allowed to watch? Is television a service for the public or is it something we should have to buy like any other product?

At the moment there is a great deal of debate going on about the management of television. Outlined below are some of the important factors that determine the type and content of programmes that we see on our screens.

BELOW
Popular TV personalities can build powerful positions within the TV industry. Oprah Winfrey has her own production company.

Funding

It takes a lot of money to make and broadcast TV programmes, so where do channels get this money from? In the UK, everyone who owns a television set must buy a TV licence. The millions of pounds generated by the sale of TV licences go to the BBC, which uses the money to produce, buy and broadcast programmes for its two TV and five radio channels.

ITV receives its money from advertising companies. These pay to have adverts broadcast in between programmes, or the name of a product mentioned at the beginning, during and at the end of a programme. Advertising agencies in turn receive money from the companies being promoted in the adverts.

ABOVE
The Daleks, which – as people from many countries will know – feature in BBC TV's *Doctor Who*. The programme has been sold by the BBC to TV companies all over the world.

So, the BBC is publicly funded (its programmes are paid for by the public, through licence fees) whereas ITV and Channel 4 are commercially funded (their programmes are paid for by businesses, through advertising). Many countries have public as well as commercial TV stations, although the way in which public TV stations are funded varies a great deal.

TV channels can also make money by selling their programmes abroad to be shown by foreign television companies, and by releasing programmes on video to be sold in shops.

Sometimes programmes are co-funded by a number of TV companies from different countries. The wildlife programme *The Trials of Life* was made by the BBC but was partly funded by the Australian channel ABC.

Satellite and cable channels are funded mainly by advertising but they also make money by charging viewers a fee to watch the programmes they broadcast. It is possible that this system may, one day, replace the television licence fee. (More about these channels in chapter six.)

Public Service Broadcasting
The BBC has always been committed to public service broadcasting. This means that as well as showing entertaining and informative programmes, the BBC channels will broadcast a number of programmes that provide a service for various communities of people.

Examples of public service broadcasts are schools programmes, foreign language course programmes, adult education programmes, religious programmes and so on.

Originally, ITV and Channel 4 were also obliged to show public service programmes. However, many of these types of programme have a very limited audience and cannot generate much money from advertising. ITV and Channel 4 are moving away from their public service commitments. As they operate along increasingly strict business lines, they are dropping programmes which will not make a profit.

BELOW How the BBC's income is shared out.

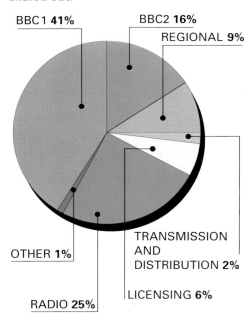

BBC1 **41%**
BBC2 **16%**
REGIONAL **9%**
OTHER **1%**
TRANSMISSION AND DISTRIBUTION **2%**
LICENSING **6%**
RADIO **25%**

Some people believe that the BBC should be a self-funded business, not a government-funded service. They believe that the licence fee should be scrapped. If this happens, it could mean the end for public service broadcasting.

Regulation and Censorship

As we have seen in chapter one, television can have a very powerful influence on its viewers. Because of this, people called censors carry out the job of regulating TV. This involves checking that programme makers and channels are following a set of rules concerning broadcasting that have been laid down by the government. They have to check the following.

- News programmes are balanced and unbiased.
- A full range of interests and subjects is covered.
- Children do not get exposed to excessive sex and violence.
- TV adverts do not break the rules that control TV advertising.

It is the role of censors to decide exactly how much sex and violence the viewer should be allowed to see. This is an area which causes a lot of disagreement. Some people believe that censorship takes away freedom - in this case, the freedom of the viewer to decide

tv and video

ABOVE TV codes dictate that news programmes should be balanced and unbiased.

what he or she wants to watch. Others believe that censorship is necessary in order to protect viewers, particularly young ones, from seeing things that might be disturbing or damaging in some way.

The BBC regulates itself. In other words, it is the job of people within the BBC to make sure that their programmes do not break government regulations. However, ITV and Channel 4 are regulated by the ITC (Independent Television Commission), which, as the name suggests, is a separate, independent body. The ITC replaced the IBA (Independent Broadcasting Authority) after the 1990 Broadcasting Act was passed.

The ITC vowed to have a 'lighter touch' than the IBA. They intended to be less strict about what broadcasters on ITV and Channel 4 can or cannot show. For example, they no longer preview programmes or make decisions about such details as programme schedules or the size of game show prizes. This could be the first step of what has come to be known as de-regulation. Some people argue that we are moving towards a TV industry that will have no regulation or restrictions; where anything goes. They believe channels will become politically biased, and there will be no balance of different types of programmes. Their main worry is that there will be fewer restrictions on programmes containing sex and violence.

media **watch**

Occasionally, governments take regulation and censorship into their own hands. They may try to ban a programme from being broadcast if they consider it to be too politically sensitive. This has happened in the UK to programmes about the IRA and the Falklands War. Sometimes, bans are overturned in court after the programme makers have successfully claimed that the government has acted unlawfully.

It is much more difficult to regulate or censor satellite channels. This is because programmes are broadcast in one country and bounced off communication satellites to dishes in a number of other countries. It has been possible to watch programmes on satellite channels in Britain that would not be allowed to be broadcast by the BBC, ITV or Channel 4 because of their sexual content.

BELOW American 'TV evangelists' – church ministers who preach to their congregations through the medium of television – are very powerful people in America. Here, Jimmy Swaggart asks viewers to forgive him for committing adultery.

tv and video

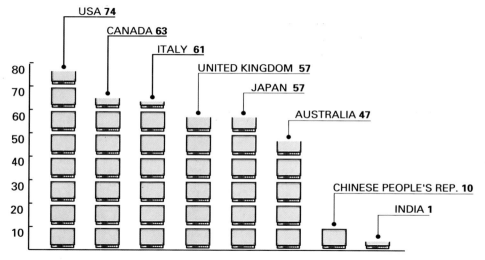

ABOVE TV sets per 100 population.

TV Watchdogs

These are organizations run by members of the public which monitor TV channels and publish reports about them. Although they do not actually have any direct power over the channels, they try to influence politicians to make the changes they want. Occasionally, they will try to influence the channels themselves. They can attempt to do this in a number of ways, for example by protesting about certain programmes.

The National Viewers and Listeners Association (NVLA) in the UK is predominantly concerned about sex and violence on television. The Voice of the Viewer and Listener (VVL) is concerned about the effects of de-regulation on the quality of programmes.

Television programmes such as *Right to Reply* and *Points of View* allow members of the public to air their views or complaints about programmes.

Ownership

Does it matter who owns television companies? It is difficult to say for sure how much influence the owner of a channel has over the programmes it produces and broadcasts. But many people think it has a very important effect on the types of programmes that appear on our screens.

We saw in chapter one how state-owned television, representing only one point of view, could have an almost brainwashing effect on its viewers. It is possible that a similar situation could arise if all the television companies in one country were owned by the same person. The channels might not feature different points of view on certain important issues. They might only show adverts which publicized the owner's other companies and products. The owner would be in a position to decide which stories were featured on news and current affairs programmes.

media **watch**

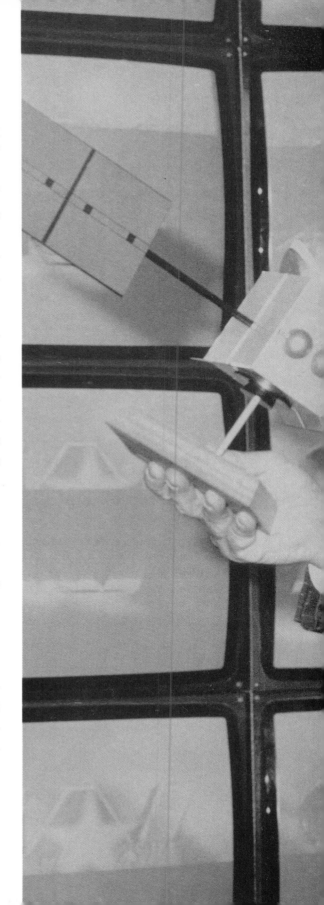

It would be against the law in the UK for one person to own all of the TV channels. However, the businessman Rupert Murdoch owns the satellite TV station Sky as well as the newspapers the *Sun* and *The Times* (among others). Some of his newspapers seem to feature adverts for his television channels, disguised as news stories.

RIGHT Rupert Murdoch with a model of the satellite which beams Sky TV on to our screens.

The owner's influence is very clear on many of the cable channels in the USA. For example, some religious channels in America aim to promote a particular religious organization and raise funds for that group.

Activities

You are the programme planner for a channel owned by one of the following.
- An environmental organization like Greenpeace.
- A high-powered business person.
- A political party.
- A religious organization.

• List the items in a half-hour news programme which reflect the views and interests of the owners. Design the station logo which appears between programmes.

media **watch**

Telly addicts:
the audience

Television needs an audience. The audience is the mass of viewers which TV programmes are aimed at, designed for and consumed by. Without a mass audience there would be no money to make and broadcast TV programmes. Television companies spend a lot of time and money on research into the audience's viewing habits.

Ratings

Television ratings are lists of figures published weekly which tell us approximately how many people were watching each programme broadcast on each channel. They are compiled by professional researchers, who select a sample of the population and attach special meters to their TV sets. Although there are only a few thousand people in the sample audience, these ratings are considered to provide an accurate picture of the nation's viewing habits as a whole.

By studying the ratings it is possible to build up some idea of what programmes had the most viewers on any given night; the number of people watching TV at various times of the day; and what types of programme are most popular.

Soap operas usually get the highest ratings, which is an indication of their importance to us. In Britain, *Coronation Street* gets an audience of 16-20 million, whereas *Brookside*'s regular

BELOW
Big Breakfast star presenter Chris Evans. The show gets good ratings figures and demonstrates the importance of choosing a presenter who will be immediately popular with the target audience.

tv and video

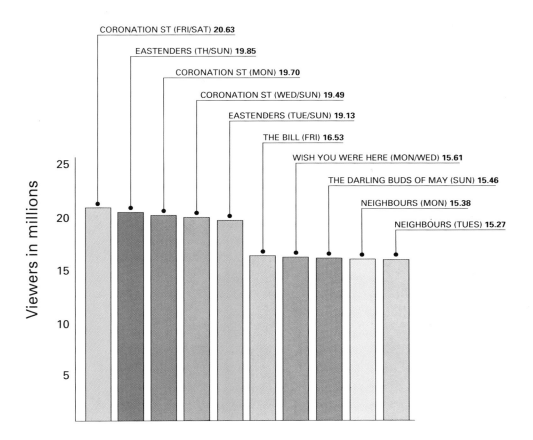

audience of 4-6 million makes it one of Channel 4's most popular programmes.

Schedules

Ratings are used when a channel works out its schedules. Schedules are the times allotted for TV programmes to be shown. A great deal of care is taken by the planners who compile schedules, in order to ensure the largest overall audience for the channel.

The planners divide each day into chunks of time, called slots, which must be filled. Most evenings, for example, are divided into half-hour slots which begin on the hour and at half past the hour. TV ratings figures show that most people watch television between 7pm and 9pm – 'prime time' as it has come to be known. Prime time slots are the most prized by programme makers, as they have the largest potential audience.

Channels can identify their most popular shows by looking at the ratings. Each channel wants to ensure that it will get the maximum possible audience for each programme. This means that channels generally show their most popular programmes, such as soap operas and situation comedies, during prime time.

ABOVE TV Ratings (figures supplied by BARB).

tv and video

When compiling television schedules, it is important not just to consider your own programmes in each slot, but also those on the other channels. Programmes on other channels are competing for viewers with your programmes. BBC 1's evening schedule is carefully arranged according to ITV's schedule and vice versa.

For example, *Eastenders* (BBC 1) and *Coronation Street* (ITV), two of the most popular programmes, are on alternate nights of the week so that they do not clash. Many people watch both programmes. There would be a risk for both channels of losing viewers if the programmes were broadcast at the same time. This informal 'agreement' between the TV companies not to compete with each other is called complementary programming.

On Saturday mornings and weekday afternoons, BBC 1 and ITV show children's programmes which directly compete with each other for the same viewers. Both channels even group these programmes together under similar names: 'Children's ITV' and 'Children's BBC'. Putting programmes that are very similar and aim to

LEFT
Coronation Street has been hugely successful for many years. Notice the edge of the studio set on the left of the picture.

33

media **watch**

ABOVE
Tony Grieg, commentator for Channel 9, Australia's main cricket channel.

attract the same type of audience on at the same time is known as counter-programming.

Channels tend to offer a different type of programme while the popular soap operas are being shown on the other side. This tactic of offering an alternative is also used when one channel shows a big sporting event such as a live football or cricket match. The other channels often show a 'big' film. This is in the hope of winning over some of the viewers and catering for people who don't like sport. The two programmes generally have a similar running time and so fill the same slots in the schedules.

Minority Interest Programmes

Not everybody likes to watch the same programmes. The television audience as a whole can be broken down into groups of viewers with particular tastes or interests. Both BBC 2 and Channel 4 were created with the intention of showing minority interest programmes. These are programmes which are considered to have a relatively small potential audience because their subject matter is specialized, or is not of interest to most people.

Examples of this type of programme include arts programmes about, for example, classical music or books; gardening programmes; unusual sports such as sumo wrestling; and magazine programmes aimed at specific ethnic groups, such as *Look East*.

However it is not always a programme's subject matter which makes it a minority interest programme. In countries where several languages are spoken, there is a need to broadcast programmes in different languages.

tv and video

This has led to the setting up of Spanish-language cable channels in the USA and French-language channels in Canada.

The Three Minute Culture

Surveys conducted in the USA about television viewing habits suggest that the average time a viewer now spends watching a programme before he or she switches to another channel is three minutes. Two of the main reasons for this are that we can change channels without having to leave our chairs and we have an increasing number of channels to choose from. (Some people in the USA are able to choose from as many as 100 channels!) Also, channels like MTV, with its back-to-back pop videos, only require an attention span of about three minutes anyway.

These viewing habits have led to the use of the phrase 'three minute culture' to describe the modern world we live in today. The phrase describes the way in which people are constantly shifting their attention from one place to another – demanding instant entertainment or information. Some people argue that this is not a bad thing but just an indication of how quickly we are now able to get information from TV.

Activities
• Compile a set of TV ratings using a sample audience of people in your school. Ask as many different people as you can what their ten favourite programmes are. Ask your parents as well.
• Compare your results with the real television ratings which you can find in some newspapers and magazines, and on teletext. From your results, decide which programmes you think appeal mainly to young people and which ones appeal more to older people.

BELOW
How popular is *Beverly Hills 90210* among people at your school?

35

media **watch**

Tuning in to the future:
new technology

Transmit, receive, decode, fast-forward, play, load, programme! The invention of several new home entertainment systems over the past twenty years has drastically changed the way we use our televisions. This new technology includes stereo systems, video recorders, camcorders, satellite receivers, cable systems, computer games consoles and so on.

Satellite

Satellite television programmes are transmitted by beaming a signal into space and reflecting it off a communication satellite. The signal is bounced back down to earth, where it can be picked up by anyone who has a satellite dish. Providing they have paid a subscription fee, the programme will be decoded for them to watch.

A TV company broadcasting in this way is able to transmit its programmes across a far wider area than is possible using earthbound transmitters. Programmes can be broadcast in many countries by satellite. The problems with satellite technology are that the pictures are prone to interference. Also, the hardware needed to receive the programmes is cumbersome and unsightly.

**BELOW
A TV satellite dish.**

**RIGHT
A communication satellite.**

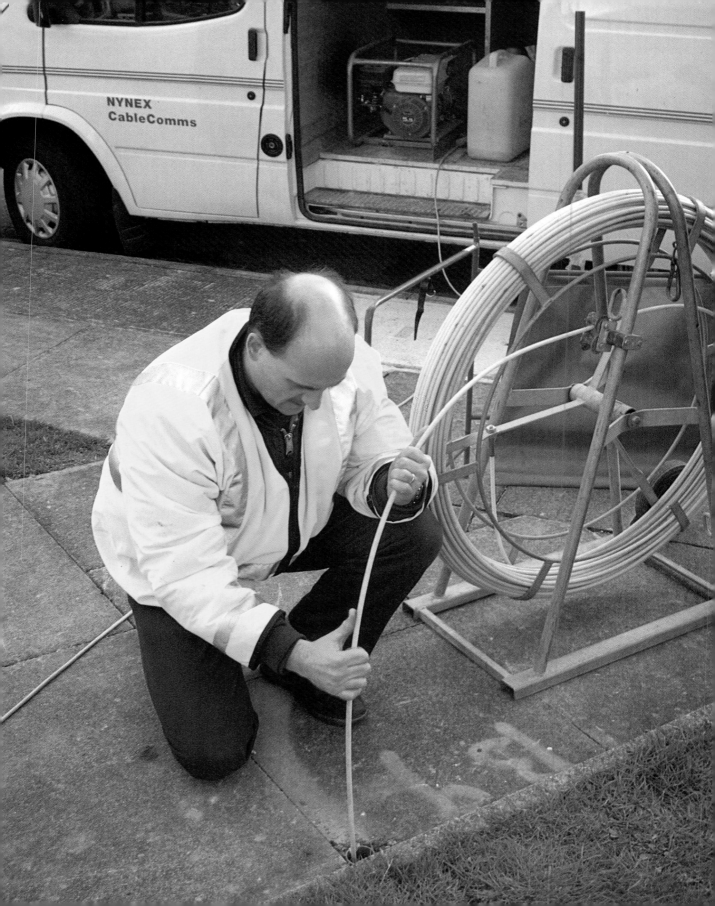

tv and video

Cable

Another way of broadcasting television channels is via underground cables similar to telephone lines. This technology has been around for many years. It was introduced as a means of relaying television in Britain and the USA as early as the 1950s. Modern fibre optic cable has the potential to carry hundreds of channels (including satellite channels) down a very fine cable, without picture interference.

Cable is very popular in parts of Europe. In Belgium and Holland, over 80% of homes are cabled. A number of different languages are spoken in these countries, and cable gives people access to TV in various tongues. In the USA, well over 40 million homes have been cabled. The disadvantages of cable are that it is very expensive to make and install.

We now have access to more television channels than ever before, but do more channels automatically mean more choice? Critics of satellite and cable channels argue that they just provide more of the same: more game shows, more chat shows, more soap operas and more repeats. These programmes are cheap and popular, so they are attractive to programmers. Satellite and cable channels are only interested in making money. They have no commitment to public service broadcasting. And they are not really interested in producing expensive 'quality' programmes, such as television dramas.

However, many people have welcomed the opportunity to have sports, news, pop music and movie channels on tap twenty-four hours a day.

LEFT TV cables being laid under a street.

BELOW NYNEX – an American cable TV station.

Video

Recording television pictures on to videotape was developed in the USA and Japan in the 1950s. It was not until 1978 that the modern domestic video recorder was launched in Europe. VCRs rose in popularity over the next decade. Today, well over a half of all households in Britain own a video recorder. In comparison, television grew in popularity far more slowly.

Video recorders let us record television programmes on to tape and then play them back and watch them later. We can also hire videos of films previously released at the cinema, or buy video tapes of anything from sport to pop music. This means that we can decide for ourselves what we want to watch and when we want to watch it.

LEFT With the amount of electrical equipment and cables needed by a TV technical crew, outside broadcasting in the 1950s was very difficult, and cameras had to remain in one place.

Through the use of video, viewers have gained more control over what they can watch. However, some people and organizations are worried about this. The NVLA in Britain is concerned that young people have access to pornographic or excessively violent videos. Members of this group say that these videos will harm young viewers. Other people believe video has freed us from the control of the censors: it is more difficult for the authorities to tell us what we can or cannot watch.

Camcorders

The word camcorder is a combination of the words camera and recorder. As the name implies, a camcorder is a video camera and a video recorder combined into one portable unit. They have become cheaper and are increasingly popular. This means that more and more people are able to watch videos on their televisions that they have made themselves.

Camcorders can be used to video subjects such as the family holiday, or to make

BELOW Today's camcorders are very easy to operate.

tv and video

programmes about social or political issues. People are using camcorders to communicate with other people about issues that are important to them. Pressure groups and political organizations can screen videos at meetings or circulate them through specialist catalogues.

Occasionally, home-made programmes are broadcast on TV. The BBC's *Video Diaries* series gives ordinary people from different walks of life the opportunity to broadcast documentaries about themselves, filmed with camcorders. Regional cable companies also broadcast programmes and footage filmed on camcorders, partly as a way of making local television cheaply. Because it gives members of the public access to TV broadcasting, this type of programming is referred to as 'access television'. Access television covering a very wide range of topics has existed for many years on some American cable channels.

Computers and Interactive TV

The future of television and video is digital. Digital information is stored in the form of millions of numbers. CDs and computer discs store digital information. Analogue information is stored by making a magnetic imprint on tape. Video or

**LEFT
With the use of paintboxes, technical staff are able to introduce all sorts of computer-generated graphics on to our screens.**

43

media **watch**

ABOVE
Modern TV technology – the big outdoor screen. Does this increase the amount of information and entertainment available to the viewer, or is it just an eyesore?

audio cassettes store analogue information. The advantage of digital sound and vision is that it can be copied without the quality deteriorating. If you have ever edited video tape, you may have noticed that the quality gets worse each time you copy it.

The other major advantage of digital technology is that we are able to interact with it. The best example of this is a computer game. If we type in a command on the keyboard or even just move a joystick, we are adding new information. What we see on the screen changes according to the commands we give. This is how most visual special effects are created for television and cinema. Images can be digitised, in other words processed so that they can be stored on digital video in the form of millions of numbers. This means that we can change the image just by changing some of the numbers. Digital video is compatible with computer hardware, so potentially the viewer can interact with it using a computer.

The Future
All this modern technology increases the viewer's editorial control. Increasingly, we can choose what we want to watch and when and how we watch it. If we do not like what is on offer we can even make our own alternative. It may not be long before we can actually interact with the television programmes we watch, changing them as we please or even entering into the action ourselves!

Activity
• Devise an interactive game which combines both computer and video technology. It may be in the form of a TV drama in which the viewer decides the outcome by using his or her computer to change what the characters do or say. It could be informative or educational and designed for a specific place or purpose such as a museum. Try to make it as imaginative as you can.

glossary

Broadcast To transmit a programme 'on air' to an audience.

Composition The arrangement of people or objects within a frame.

Current affairs Up-to-date news stories.

Decode To make clear a scrambled signal.

De-regulation The relaxing of rules and regulations.

Edit To prepare a TV programme or film for broadcast by cutting and arranging the filmed material in the order in which it is to be shown.

Fibre optic cables Tiny cables which can carry masses of information.

Formula A method for creating a certain style of programme. A formula is like a recipe for a TV programme i.e. it should contain key ingredients.

Frame Everything that is in view of the camera.

Location A place where a scene is filmed outside a studio.

Logo A symbol or emblem of a company.

Politically sensitive programme A TV programme which contains information that a government does not wish to be made public, or is embarassing to the government, or is about a controversial subject or personality.

Regulate To bring TV programmes into line with certain rules and guidelines.

Re-take The re-filming of a shot.

Satellite An object in space, orbiting the earth, used as a transmitter.

Scanner Moving beams of light which create a picture.

Shot A continuous piece of action, on film or video, uninterrupted by an edit.

Transmit To send a radio or television signal.

further reading

Manuel Alvarado, *Television and Video* (Wayland, 1987)

Patrick Barwide and Andrew Ehrenberg, *Television And Its Audience* (Sage, 1988)

BFI Yearbook 1990 - 1991

Mike Clarke, *Teaching Popular Television* (Heinemann Educational Books, 1987)

Brian Dutton with John Mundy, *Media Studies: An Introduction* (Longman, 1989)

Sue & Wink Hackman, *Television Studies* (Hodder & Stoughton, 1988)

Stephen Kruger & Ian Wall, *The Media Pack* (Macmillan Education, 1987)

Conrad Lodziak, *The Power Of Television* (Frances Pinter, 1991)

National Museum of Photography, Film and Television, *Television - The First Fifty Years*

Jane Root, *Open The Box* (Comedia Publishing Group, 1986)

Cameron Slater, *Planning The Schedules* (Hodder & Stoughton, 1987)

notes for teachers

TV and Video can be used in a variety of ways to meet the requirements of the English National Curriculum. The examples of TV programmes in their historical and social settings give opportunities for interpretation and comparison of media texts. The book provides material as a basis for three approaches to media:

i) Media languages. How do visual texts produce meaning?
ii) Representation. How are individuals and social groups portrayed in television?
iii) Producers and audiences. Who produces texts and why? How do audiences respond to them?

The activities are directly related to the National Curriculum, levels 6, 7 and 8, and are designed to develop critical skills and creative powers through analysis and production of media artefacts. Concepts such as purpose, audience and medium are strongly emphasized. In addition, this practical and analytical work will involve negotiation, problem-solving, group decision-making, selection and editing, as well as developing communication skills.

TV and Video and other books in the Media Watch series will also provide valuable material for students preparing for GCSE Media Studies and GCSE Communication.

index numbers in **bold** refer to captions

ABC 23
access television 43
actors 19,20
adult education programmes 24
advertising 22,23,24
Australia 8,**34**

Baird, John Logie 5,6
Barr, Roseanne **11**
BBC 5,6,7,22,23,24,25,26, 33,43
Belgium 39

Beverly Hills 90210 **35**
BSkyB 28,**29**
budgets 10,16,18

cable TV 23,28,35,**38**,39, 43
camcorders 36,41-3
cameras 19,20,**21**,**40**
Canada 35
cartoons **10**,11
celebrities 14,18
censorship 9,24,26
chat shows 31
Cheers 11

children's TV programmes 10
China 9
colour broadcast 6,7
commercial TV 23
complementary programming 33
Coronation Street 30,**32**, 33
Cosby, Bill 18
cost of producing TV programmes 10
costumes **16**,19,21
counter-programming 33

media **watch**

de-regulation 25,26
designers **18**,19,21
directors 19,20
Doctor Who **23**
documentaries 10
drama 10,16,19,39

Eastenders 33
editing 7,21

film crews 19,20
Flintstone, Fred **10**
Forsyth, Bruce **15**
funding of TV 22

game shows 10,14,39
genre 10-15
Germany 6

history of TV 5-9
Holland 39
House of Eliott **16**

India 7
interactive TV 43-4
ITC (Independent Television Commission) 25
ITV 6,7,22,23,24,25,26,33

Japan 41

licences 22,23,24
lighting 20
live recording 20
location filming 16,**20**,21

minority interest programmes 34
MTV 35
Murdoch, Rupert 28,**29**

Neighbours 11
news and current affairs programmes 12,**13**,14,24

balance and bias in 24,25
NVLA (National Viewers and Listeners Association 27,41

ownership of TV 27-9

presenters **5**,10,12,14,15 **30**
prime time 31
production 16-21
  post-production 21
  pre-production 16-20
props 19,20
public service broadcasting 23,24,39
public TV 23

ratings 30,31,35
recording 11,19
regulation of TV 24-6
rehearsing 20
religious programmes 24
religious TV channels **26**,28
repeats 39
research into viewing habits 30-3

satellite TV 7,8,26,**28-9**, **37**,39
  dishes 36
satellite TV stations 23, 26,36,39
  regulation of 26
  quality of programmes produced by 39
scenes 11,19,21
schedules 21,25,31,33,34
schools programmes 24
scripts 10,15,18,20
  script editors 18
sets 10,14,15,19,20,**32-3**
sex on TV 24,25,26
  on video 41

situation comedy 10,11, 16,20,31
soap opera 10,11,30,31, 34
sports programmes 16, 34,39
storyboards 19,20,21
studios 16,20,21

theme music 10,12,15,19
Television
  and the global village 8
  and politics 7,8,9,12,25 26
  and the three minute culture 35
  as educational tool 7
  impact on cinema audiences of 6-7
TV production companies 16,18,**22**
TV technology 36-44
TV watchdogs 27

USA 6,**7**,8,**9**,11,**26**,28,35 39,41

video 5,7,9,20,21,23,41
violence on TV 8,**9**,24,25
  on video 41
VVL (Voice of the Viewer and Listener) 27

Winfrey, Oprah 18,**22**
writers 18,20